YOUR KNOWLEDGE HAS VALUE

Kopal Arora

A Literature Review: Effect of Climate Change on Tropical Cyclones

Effect of Climate Change on Tropical Cyclones

GRIN Verlag

Bibliografische Information der Deutschen Nationalbibliothek:

Die Deutsche Bibliothek verzeichnet diese Publikation in der Deutschen National-
bibliografie; detaillierte bibliografische Daten sind im Internet über http://dnb.d-
nb.de/ abrufbar.

Imprint:

Copyright © 2011 GRIN Verlag GmbH
Druck und Bindung: Books on Demand GmbH, Norderstedt Germany
ISBN: 978-3-656-66986-9

This book at GRIN:

http://www.grin.com/en/e-book/274134/a-literature-review-effect-of-climate-change-
on-tropical-cyclones

GRIN - Your knowledge has value

Der GRIN Verlag publiziert seit 1998 wissenschaftliche Arbeiten von Studenten, Hochschullehrern und anderen Akademikern als eBook und gedrucktes Buch. Die Verlagswebsite www.grin.com ist die ideale Plattform zur Veröffentlichung von Hausarbeiten, Abschlussarbeiten, wissenschaftlichen Aufsätzen, Dissertationen und Fachbüchern.

Visit us on the internet:

http://www.grin.com/

http://www.facebook.com/grincom

http://www.twitter.com/grin_com

Literature Review

Effect of Climate Change on Tropical Cyclones

Kopal Arora

Inroduction:

Since tropical cyclones(TCs) are one of the major geophysical cause of loss of life and property, it is important to understand if there is any change in the frequency and intensity of TCs due to anthropogenic climate change.
IPCC considers 0.25-0.5 C increase in warming over tropical oceans over the past few decades due to increase in greenhouse gas concentration over past 50years. During 6th International Workshop on Tropical Cyclones, a statement was released on the connection between the TCs and anthropogenic climate change. The statement was in response to the increase in number of recent high-impact TC events which includes, 10 land falling Tcs in Japan in 2004, 5 TCs affecting the Cook island in a five week period during 2005, cyclone Gafilo in Madagascar in 2004, cyclone Larry in Australia in 2006, typhoon Saomai in China in 2006, and violently active Atlantic TC season during the period of 2004 to 2005, including the catastrophic socio-economic impact of Hurricane Katrina. A few recent articles [1] have noted a large increase in TC's intensity, frequency and wind-speeds in some regions during past 5 decades, which could be attributed to the increase in the concentration of green house gases in past 50years. However, other studies explain this noticed increase as a result of better observations made and instruments used, making it easier to detect TCs. Consensus statement by the International workshop on TC-6 reported uncertain conclusions about the influence of climate change on TC after taking into account evidence both for and against.

It was concluded that no TC could be solely attributed to the anthropogenic climate change. Model and theory predicts 3-5% increase in wind speed per degree C increase in SST. But, there is inconsistency between the small change in wind speed projected by theory and modelling versus large variations reported by some observational studies. Significant limitation of measurements over some regions make detection of trends difficult. It was suggested that if increase in SST continues, susceptibility to TC storm surge flooding would strengthen.

Main Part:

Here we begin with a few papers published in support of the argument and then present unfavoured arguments by other authors. Intensity of a cyclone measure its destructive potential. Emanuel [2] in 1987 used a simple Carnot cycle model to measure the maximum intensity of a hurricane at higher temperature which he assumed as a result of CO_2 increase in the atmosphere. Using GCM with twice the present amount of CO_2, he predicted 40-50% increase in the destructive potential of the cyclones. In 1996, Nicholls et. al. proved downward trends in the frequency of intense Atlantic hurricanes during the past five decades. Theory(K.Emanuel-1987) [2] and Modelling(Knutson and Tuleya-2004) [3] anticipate increase in TC intensity with global warming. Emanuel(2005) [4] describes power dissipation index(PDI) which

depend upon the storm intensity and lifetime. The author shows that the index has increased remarkably since the mid-1970s.

The index is shown to be very well correlated to the SST, reflecting climate signals including Multi- decadal oscillations in the north Atlantic and north pacific and global warming. His results indicates a rising trend in tropical cyclone destructive potential and significant increase in the hurricane related losses owing to the increase in coastal population.

According to IPCC 2001 report [5], the increase in droughts, TCs, and extreme high tides is probable at confidence level greater than 66% To understand the effect of model used in simulation, Knutson and Tulya in 2004 performed 1300 five day idealized simulations using high resolution versions of GFDL hurricane prediction models. After assessing 4 different moist convection parametrization and no convection parametrization in hurricane models, they concluded net increase in the storm intensity and near storm precipitation rates. The fractional change in precipitation was found to be more sensitive to the parametrization. Convective available potential energy(CAPE) was found to be 21% higher for more CO_2 amount in the atmosphere. On this basis, he concluded that even if the cyclone frequency remain constant with time, one should expect increase in the occurrence of highly intense storms, of category 5. According to Kevin Walsh [9], considering reliability of the model predictions, more intense cyclones can be detected in Atlantic only after 2050. A study made by Emanuel in 2007 indicated regional affect of climate change on TCs. Revised estimation of KE of the TCs in Atlantic and western North pacific was performed and it was found that Atlantic variability on time scales of few years or more is significantly correlated with tropical Atlantic SST while in the western north Pacific, this correlation was considerably weaker. Using basic theory and empirical statistical analysis it was shown that much of the fluctuation in the ocean basins can be explained by variations in potential intensity, low level vorticity and vertical wind shear.

In the Atlantic the potential intensity, low level vorticity, and vertical wind shear strongly covary and are highly correlated with SST unlike, Pacific where the three factors were weakly correlated. Publications against the view are as follows. A couple of studies performed on the TCs frequency detect absence of trends in their frequency(Landsea, Nichollas, & Gray-1996 and Chan & Shi -1996). Although some statistical methods predict more than 300% increase in TC intensity in Atlantic by the late 21st century, existing downscaling models or alternate statistical models does not support this dramatic variation(Vecchi et. al 2008). Atlantic, Solow and Moore (2002) showed absence of any periodicity in the N.Atlantic hurricane activity from 1900 to 1998. Nicholls et al. (1998) indicated that there has been downward trend in TC numbers in the Australian region since the late 1960. In the northwest Pacific, Chan and Shi (1996) found a downward trend in the occurrence from 1959 to late 1980s, and there has been upward trend since then. No trend in the South Indian ocean (Henderson-Sellers et al. 1998) but a downward trend in the North Indian ocean (Raghavan and Rajesh 2003).The paper reviews present understanding of the influence of climate change on tropical cyclones. According to the author, there has been no change in the region of cyclone formation; however, there has been general agreement about the change in tracks and frequency. Changes in intensity should be detectable in Atlantic sometime after 2050.

Paleotempestology (Liu and Fearn 1993,2000, Donnelly et al. 2001a,b), which focus on the analysis of paleoclimate information of past hurricanes to extend our present understanding for a longer period than the observed one, has been involved in the study of climate change effects on tropical cyclones and it shows substantial change in incidence over the period of centuries and millennia.

Conclusions:
Considering the results discussed so far, it is apparent that the intensity of tropical cyclones will increase in future with slight decrease in their frequency. The PDI index which is the measure of the destructive potential of a tropical cyclone seems strongly related to the SST and tropical cyclone intensity over the Atlantic ocean but not so in west pacific.

Effect of Climate change on tropical cyclones over Indian ocean:

TC Frequency over the North and South Indian ocean has been used as a measurement parameter to see any influence of climate change on TC but the intensity, the Power Dissipation Index, over the region has not been taken into account. Further work could be done by considering the intensity/PDI of TC over Indian ocean and checking its variability against climate change parameters for instance, SST and low level vorticity.
Arabian sea and the bay of Bengal constituting Indian ocean are of considerable importance, since the coastal regions in the vicinity of the Indian ocean is vulnerable to tropical cyclones, and the areas are densely populated. Most importantly the Indian ocean is characterised by statistically significant surface warming trends. This property is more prominent in Indian ocean then in North Pacific and North Atlantic ocean [8].

Thus, if climate change have considerable effect on cyclone formation, Indian ocean could be more prone to these extreme events than any other ocean and signature of the influence of increasing temperature due to climate change on tropical cyclone would first be apparent in Indian ocean.

Not more than 5 to 6 instabilities over North Indian ocean reach the tropical storm stage [13]. However, discussions by Webster et al. [19] showed significant increase in category 4 and 5 storms. On the contrary, Landsea et. al.(2006) [20], showed that the used databases were not reliable. Encouraged by a paper on climatology of Atlantic Hurricanes by Landsea CW (1993) [21], Karl Hoarau and Ludovic Chalonge [6], re-analysed cyclone intensity from the polar and geostationary satellite pictures with the Dvorak's technique (based enhanced infrared technique) found that the decadel extreme intensity remain almost unchanged during. But, North Indian Ocean has potential to intensify the cyclones to about 160 - 170 knots. A study made by Vidale et al(2009) [7] compared three state of art GCMs and found that the model resolution is not important in representing the climatoloty of storm track density, however can help in detecting the intensities of cyclones. Moreover, high resolution of more than 50 to 60Km is required to determine the category 4 and 5 cyclones and realistically represent the full spectrum of tropical cyclone intensities. A climate change impact model has been used by Alam and Ahmed(2009) for Bangladesh. Their analysis suggests Bangladesh is expected to be a major victim of climate change. A study by F.Ren et. al. [11] showed downward trend in tropical cyclone precipitation, TC frequency and typhoons that affect China during 1957 - 2004. This decrease in number could be associated with increase in more intense cyclones.

Indian ocean contributes to 5% of the total tropical cyclones. For every four cyclones produced per year, three forms in Bay of Bengal and one in Arabian sea. Tropical cyclone Intensity simulation shows an increase in the frequency of more intense cyclones during past three decades [13]. Model simulations show an overall increase in wind speed of cyclones during May, October, and November in Indian ocean. Thus, there is an apparent increase in the number of stronger tropical cyclones (Maximum wind speed of 96 Knots and above).

Conclusions:

From the discussion above we may infer that different oceans across the globe have dissimilar response to the climate change which may be due to their unique characteristics in terms of differential heating, geographical location, vicinity of land surrounding them and internal ocean topography.

The first graph below shows the increase trend in the potential destructive intensity(PDI) of the tropical cyclone with the rising SST over Atlantic Ocean. Similarly, the second plot representing the rising progression of PDI with the SST over West Pacific Ocean.

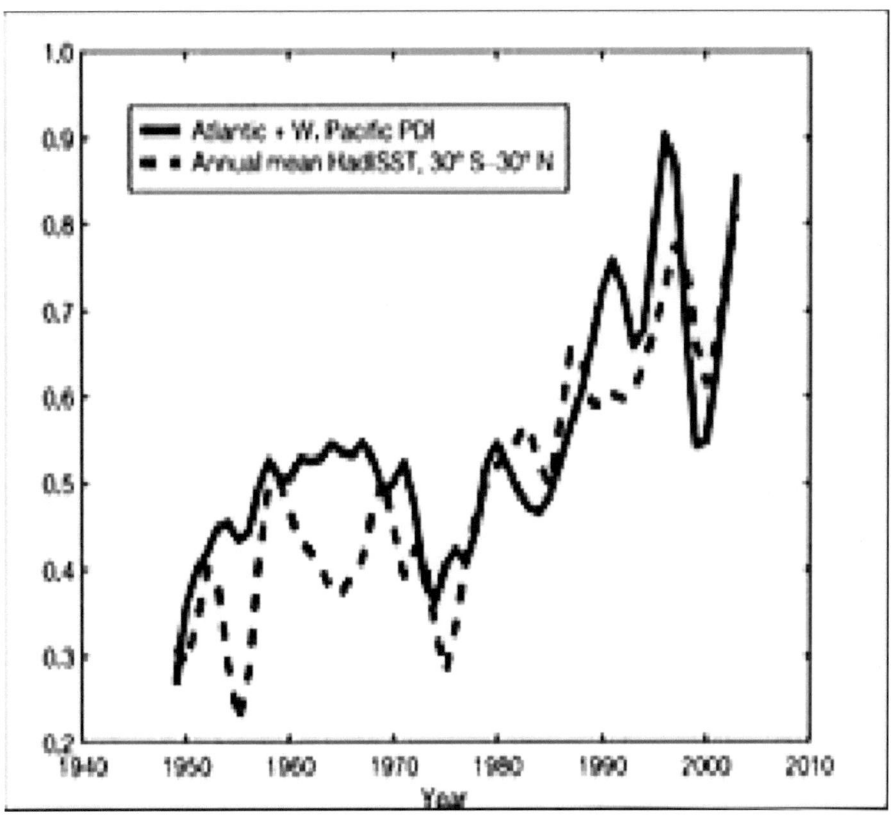

Source: Emanuel , K. A., 2005: Increasing destructiveness of tropical cyclones over the past 30 years. *Nature*, **436**, 686-688

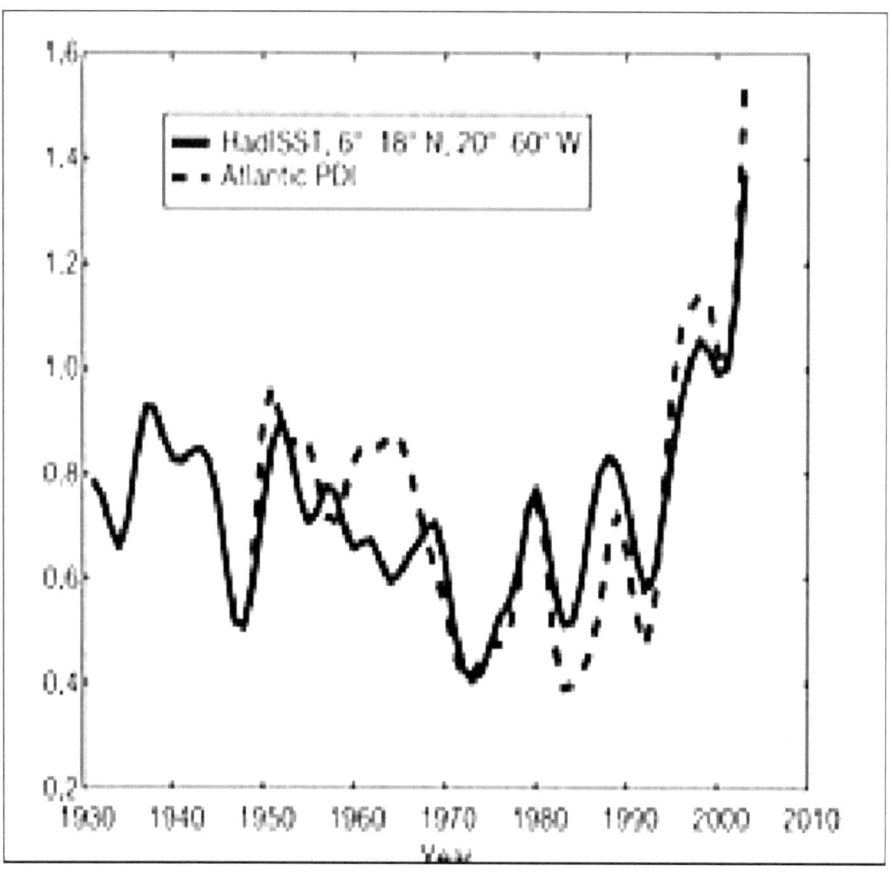

Source: Emanuel , K. A., 2005: Increasing destructiveness of tropical cyclones over the past 30 years. *Nature*, **436**, 686-688

The decreasing frequency of TC and increasing intensity can be inferred form the graph below which shows increasing number of highly destructive cyclones, category 4 and 5.

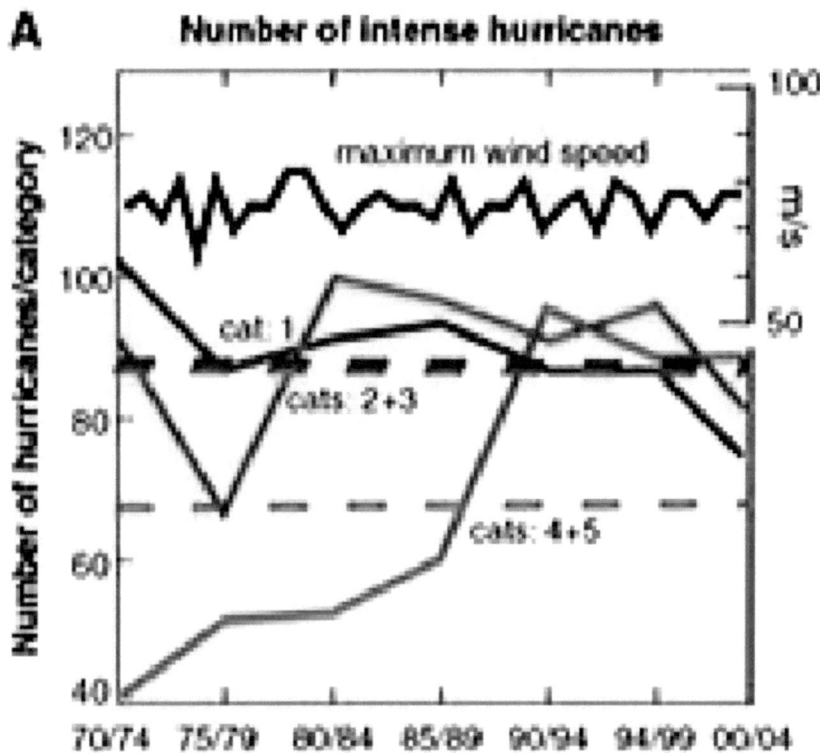

A **Number of intense hurricanes**

Source: Schade, L. R. and K. A. Emanuel, 1999: The ocean's effect on the intensity of tropical cyclones: Results from a simple coupled atmosphere- ocean model. *J. Atmos. Sci.*, **56**, 642-651

References:

[1] Emanuel, K. A(1988) .: The maximum intensity of hurricanes. J. Atmos. Sci.,45, 11431155,Emanuel, K. A.,(1995): Sensitivity of tropical cyclones to surface exchange coefficients and a revised steady-state model incorporating eye dynamics. J. Atmos. Sci.,52, 39693976.,Holland, G.J. (1997),: The maximum potential intensity of tropical cyclones. J. Atmos. Sci.,54, 25192541(1997)

[2] Emanuel, K. A (1987): The dependence of hurricane intensity on climate. Nature 326, 483485.

[3] Knutson, T. R & Tuleya, R. E (2004): Impact of CO2-induced warming on simulated hurricane intensity and precipitation: Sensitivity to the choice of climate model and convective parameterization. J. Clim. 17, 3477 − 3495.

[4] Emanuel,K. (2005) Increasing destructiveness of tropical cyclones over the past 30 years, Nature, 436, 686688.

[5] IPCC (Intergovernmental Panel on Climate Change) Houghton JT, Ding Y, Griggs DJ, Noguer M, van der Linden PJ, Xiaosu, D (eds) (2001) Climate change 2001: the scientific basis. Contribution of Working Group I to the Third Assessment Report of the Intergov-ernmental Panel on Climate Change (IPCC). Cambridge University Press,Cambridge and New York

[6] Karl Hoarau and Ludovic Chalonge: "A Climatology of Intense Tropi-cal Cyclones in the North Indian Ocean Over the Past Three Decades (1980 - 2008)"

[7] P.L. Vidale, M. Roberts, K. Hodges, J. Strachan, M.E. Demory, and J. Slingo (8-11M arch2009): "Tropical Cyclones in a Hieararchy of Cli-mate Models of Increasing Resolution", International Conference on Indian Ocean Tropical Cyclones and Climate Change(IOTCC), Mus-cat, .

[8] Knutson TR, Delworth TL, Dixon KW, Held IM, Lu J, Ra-maswamy V, Schwarzkopf D, Stenchikov G, Stouffer RJ (2006) As-sessment of twentieth-century regional surface temperature trends using the GFDL CM2 coupled models. J Climate 19(9):16241651. http://www.gfdl.noaa.gov/ reference/bibliography/2006/tk0601.pdf

[9] Kevin Walsh (4 August 2004):Tropical cyclones and climate change: unresolved issues,CLIMATE RESEARCH Clim Res, Vol. 27: 7783

[10] M.J.B. Alam and F. Ahmed (8-11 March 2009): "Modeling Climate Change: Perspective and Applications in the Context of Bangladesh", (IOTCC), Muscat.

[11] Fumin Ren, Guoxiong Wu, Xiaoling Wang, and Yongmei Wang (8-11 March 2009):"Changes in Tropical Cyclone Precipitation Over China"(IOTCC), Muscat.

[12] Thomas R. Knutson (8-11M arch2009): "Tropical Cyclones and Climate Change: An Indian Ocean Perspective", (IOTCC), Muscat.

[13] O.P. Singh (8-11M arch2009): "Recent Trends in Tropical Cyclone Ac-tivity in the North Indian Ocean", (IOTCC), Muscat.

[14] Mohammed Haggag, Takao Yamashita, Kyeong Ok Kim, and Han Soo Lee (8-11M arch2009): " Simulation of the North Indian Ocean Tropical

Cyclones Using the Regional Environment Simulator: Application to Cyclone Nargis in 2008", (IOTCC), Muscat.

[15] Ajit Tyagi, B.K. Bandyopadhyay, and M. Mohapatra (8-11M arch2009): "Monitoring and Prediction of Cyclonic Disturbances Over North Indian Ocean by Regional Specialised Meteorological Centre, New Delhi (India): Problems and Prospective",(IOTCC), Muscat.

[16] Y.V. Rama Rao, A. Madhu Latha, and P. Suneetha (8-11M arch2009): "Evaluation of the WRF and Quasi-Lagrangian Model (QLM) for Cyclone Track Prediction Over Bay of Bengal and Arabian Sea",(IOTCC), Muscat.

[17] Krishna K. Osuri, A. Routray, U.C. Mohanty, and Makarand A. Kulkarni (8-11M arch2009): "Simulation of Tropical Cyclones Over Indian Seas: Data Impact Study Using WRF-Var Assimilation System",(IOTCC), Muscat.

[18] R. Montroty, F. Rabier, S. Westrelin, G. Faure, Lok Berre, and Laure Raynaud (8-11M arch2009): "Impact of Rain-Affected SSM/I Data Assimilation on the Analyses and Forecasts of Tropical Cyclones, and Study of Flow-Dependent Ensemble Background Errors, Over the Southwest Indian Ocean",(IOTCC), Muscat.

[19] Webster PJ, Holland GJ, Curry JA, Chang HR (2005) Changes in tropical cyclone number, duration, and intensity in a warming environment. Science 309:18441846

[20] Landsea CW, Harper BC, Hoarau K, Knaff JA (2006) Can we detect trends in extreme tropical cyclones Science 313:452454

[21] Landsea CW (1993) A climatology of intense (or major) Atlantic hurricanes. Am Meteor Society 121:17031713